FIFTY THINGS

THINGS

TO DO IN THE

SNOW

Pavilion
An imprint of HarperCollins*Publishers* Ltd
1 London Bridge Street
London SE1 9GF

www.harpercollins.co.uk

HarperCollins*Publishers*
Macken House, 39/40 Mayor Street Upper
Dublin 1, D01 C9W8
Ireland

10 9 8 7 6 5 4 3 2 1

First published in Great Britain by
Pavilion, an imprint of HarperCollins*Publishers*
Ltd 2023

A catalogue record for this book is available from
the British Library.

ISBN 978-1-911682-57-8

This book is produced from independently
certified FSC™ paper to ensure responsible
forest management.

For more information visit:
www.harpercollins.co.uk/green

Publishing Director: Stephanie Milner
Commissioning Editor: Lucy Smith
Managing Editor: Clare Double
Design Manager: Laura Russell
Senior Designer: Alice Kennedy-Owen
Layout Designer: Hannah Naughton
Production Controller: Louis Harvey

Printed and bound by Oriental Press, Dubai

IMPORTANT SAFETY NOTICE

This book includes activities and projects that inherently include the risk of injury or damage. We
cannot guarantee that following the activities or projects in this book is safe for everyone. For this
reason, this book is sold without warranties or guarantees of any kind, expressed or implied, and the
publisher and the author disclaim any liability for injuries, losses and damages caused in any way by
the content of this book. The publisher and author urge the reader to thoroughly review each activity
and to understand the use of all tools before beginning any project. Always check that you have
permission to use the land where the activities or projects take place. Never underestimate the risk
of treading on thin ice. Children should always be supervised by an adult when undertaking any
activity or project in this book.

FIFTY
THINGS
TO DO IN THE
SNOW

Richard Skrein

Illustrations by Maria Nilsson

PAVILION

Contents

Introduction

We look up toward the heavens. Our eyes alight on that cloud, painted across the winter sky. Mainly made up of air, but also billions of droplets of water vapour, liquid water and ice crystals. There's more, too; ash, pollen, dirt, salt and even living bacteria. It is on these particles that a snowflake is born.

Water vapour collects on a particle and a tiny frozen ball is formed. This grows into a hexagonal ice crystal before branches emerge in more and more detail. The unique shape of a snowflake is dictated by its path; as it descends through wetter and colder parts of the cloud, it grows in size and weight, drawn downwards by gravity. Eventually its weight becomes too much, the trick of suspension over, and the snowflake leaves its cloud.

Its descent can take up to an hour from the cloud to the surface of the earth. It's night-time and this snowflake is part of the vanguard, the first to fall this winter. Eventually we awake in our beds and know something is different; there is a magical stillness, a different quality of silence, a distinct light in the room.

'It's snowed!'

Whether you are seeing this with the child-like excitement
of the first rare snow in warmer climates or the battle-
hardened glare of those who live with it all winter, the sense
of transformation is undeniable. The world is changed with a
blanket of snow … and the imagination is sparked.

This book is an invitation to the many ways that we can interact
with and deepen our relationship with the snowy landscape. It is
an invitation to connect with the joy and wonder of the crunch
underfoot, the snowball thrown, the quiet stillness.

This book is also something of a love letter to winter itself, and
its mystery and magic. Its stillness and power.

The snow awaits, let's head out there!

Staying safe

We are magnetically drawn to explore, enjoy and appreciate the snow. It's also important that we respect the power of winter, taking care of our companions and ourselves. Consider the following precautions:

- Plan your route and equipment needed (see pages 9–19).

- Select a route, terrain and activities appropriate to your experience level.

- Consider hiring a guide if needed.

- Be aware of avalanche dangers in your area.

- Leave information with a responsible person about your route, plans and arrangements.

- Travel in groups of at least three where possible.

- Sign guestbooks in bothies, trailheads and visitor centres with your name, date and time.

- In an emergency situation, stay calm and don't panic (if you can control only one thing, make it yourself!).

Ecological impact

As well as looking after ourselves and each other, let's also look after the environment by minimising damage, carrying our litter away and leaving no trace as much as possible. When taking or using natural materials, consider asking for permission or giving something back in return, and only take what you need.

Equipment and clothing

Whether heading out for an afternoon or a longer winter trek, it's vital to pack the right kit. On the following pages we will explore the items you may wish to bring. Not all of this kit will be needed for a local day out – you can pick and choose what you need, but do make sure you have enough to keep safe in case you are out longer than expected.

Use a good-quality backpack and make sure the straps are adjusted to fit your body correctly. It's wise to keep your gear as compartmentalised as possible in your backpack by using stuff sacks or similar. You will notice that I have not included gear for mountaineering or ice-climbing; if you expect to do either of these activities, research your destination and pack accordingly.

It's also important to remember that equipment is no substitute for knowledge, experience and the correct mindset – these are more important than any kit you can buy.

Knife

In many ways the most important tool you will own. Used for carving, whittling, stripping, scraping, cutting and maybe even making sandwiches when you forget your normal knife (although do clean it first). Please be vigilant and responsible about locking your tools away when they are not in use and do check local laws regarding the carrying of blades where you are.

Saw

Used for sawing wood and ice blocks for construction. For ice blocks, a dedicated snow saw or even a simple wood saw will work well.

In colder environments, you will go through a lot more firewood than you may expect, so a high-quality lightweight folding saw will help you to conserve energy when processing large amounts.

One neat trick that could come in useful is to employ the crook of your knee as a vice, as shown here.

Axe

A small axe fits in or onto your backpack and will be used for splitting and stripping wood, as well as downing smaller standing dead trees.

Snow shovel

Like a small axe, a shovel can fit in or onto your backpack, especially if it dismantles or is telescopic. This will be an important tool for digging shelters, and has many other applications.

Multitool

A single tool containing a knife, a file, a screwdriver, pliers, a wire cutter, a bottle opener and more? Yes, please! While it is, of course, not a substitute for other tools, it is a very handy one to have around.

Cordage

Carry at least 10 metres of paracord per person, plus extra for craft projects (such as the sledge on page 22). Consider also a safety rope for longer or more dangerous journeys, fishing line, or even making your own natural cordage!

MICROSPIKES

CRAMPONS

Snowshoes and skis

These are passports to traverse and enjoy deeper snowy terrain that would be impossible to cross in boots. Of course, take trekking poles too, which also come in handy for the air vent in your snow shelters (see page 108). To make your own snowshoes, turn to page 96.

Microspikes and crampons

Microspikes are lightweight spiked chains that can be slipped over your boots to help you traverse relatively flat icy areas. For steeper icy slopes, you may need crampons, which also attach to your footwear, but have larger spikes on the base and front.

Snow probe

This will allow you to gauge the depth of the snow around you before you start digging those shelters, and can also be used for searching for people in the event of an avalanche.

Ice axe

An important safety tool if climbing or crossing ice (see page 113), an ice axe can be used as a point of balance, for climbing or stopping yourself when sliding down an icy slope.

Candles

Pack candles to illuminate your snow shelters. As well as providing that warm glow, a flickering or extinguished candle can also be a lifesaving indicator that oxygen levels are low.

Throw line

Carry a throw line (or other length of safety rope) if you expect to be near bodies of water. It could be a lifesaver.

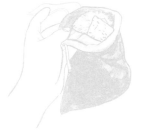

Fire-lighting equipment

I favour a ferro rod (rod of ferrocerium) for starting fires as they can withstand wet and won't run out without warning. An extra-long rod can be easier with gloves. Also carry two other forms of fire-lighters, such as a windproof storm lighter, matches or a flint and steel set.

Tinder pouch

Used to store any useful tinder you find and collect, as explored on page 121.

Reflector blanket

These are important for keeping people warm in an emergency situation and also as a heat reflector by a fire, for signalling or marking your location. Consider carrying multiple reflector blankets for longer trips.

Sun cream

Vital in all weather. Consider a lip balm, too.

First aid kit

Carry a first aid kit in a waterproof bag. Proper first aid training is also essential for people leading groups out into the wild.

Hand/body warmers

These come in handy for warming up people's extremities. You can also use hot stones, as explained on page 134.

Whistle

In an emergency situation, the sound of a whistle will travel far further than the human voice.

Binoculars

Keep binoculars on you for that magical encounter with the more-than-human world, whether a soaring bird in the distance or the intricate structure of lichen up close.

Map and compass

See page 100 for tips on using these, and consider carrying a GPS system, too.

Phone

While one of the joys of heading out into the natural world is getting away from the many screens in our lives, a phone and extra battery pack can be a lifesaver.

Headtorch

Don't forget some spare batteries!

Sleeping bag

A down sleeping bag that can protect you from the night-time temperatures in your destination is a must. Remember, a sleeping bag will conserve heat rather than create it, so get nice and hot before getting in by doing some rigorous movement like jumping jacks. If you have cold feet, massage them for 20 minutes before bedding down for the night.

Groundsheet and/or tarp

A tarp can be used to cover your shelter, as outlined on page 109, as well as act as a groundsheet under your sleeping set-up. Cheap tarps and groundsheets are great for collecting snow or other materials before pulling them along, as well as for laying out your kit in the snow.

Sleeping mat

Inflatable sleeping mats are great until they get a puncture, so you may consider a folding or rolling foam mat. Or a mat puncture repair kit!

Water bottle

You should be drinking at least 3 litres of water every day on a winter trek. See page 116 for ideas on finding water, and consider carrying a water filter in case you are unsure of the cleanliness of your water.

Cooking equipment

Lightweight and packable cooking equipment is important. If possible, take two small cooking pots, so that you can have one for water to boil in constantly and stay hydrated with teas. Take a small stove, a metal cup and bowl, and don't forget a small washing-up kit. Pack a metal cup for boiling directly on the fire or a wooden one for the aesthetics and feel of it.

Food

A friend of mine once said, 'the best thing about being in the outdoors is that you get to wear fancy dress and eat party food!' By this they meant the joy of putting on the specialist clothing and the licence to eat dense and energy-rich snacks and meals. Consider taking supplies for an extra day in case the trip takes longer than expected, and be sure to keep food in a sealable bag.

Wire and gaffer tape

A small roll of wire or gaffer tape has so many uses! It always comes in handy.

Microfibre cloth

Useful for wiping down tools and kit before they go back in your bag.

Clothing

In the outdoors in general, it's a good rule to 'dress like an onion', with layers that can be taken off and on as needed:

Mid layer of non-cotton fleece, shirt or jumper to ventilate and insulate

Base layer of non-cotton thermal underwear, long-sleeved and full-leg to keep the skin dry and warm

Reinforcement layer of a hooded down jacket to maintain body heat in the coldest temperatures

Outer layer of waterproof shell jacket and trousers to protect from the wind and weather, as well as add further insulation and ventilation

Gaiters

These will stop snow getting into the gap between your trousers and boots. You don't need a very expensive pair, but a waterproof fabric such as Gore-Tex will work well. Some waterproof trousers come with in-built gaiters. There are also some wonderful traditional versions such as the Sami shoe band, made of wool and incredibly effective.

Gloves

It's useful to take a range of gloves with you on a winter trip; wool inner gloves, durable leather outer gloves and waterproof shell gloves for handling snow and digging. Keep your gloves inside your jacket when you're not wearing them so they stay warm and soft and don't freeze wet.

Sunglasses and ski goggles

Protect your eyes from the sun with sunglasses, and pack some ski goggles for windier times.

Boots

Keeping your feet warm and dry is incredibly important for comfort and survival in the winter landscape. Everyone has their own preference when it comes to boot style and design, so try them on and see what works for you. Pair them with good-quality cold-weather wool socks, and consider waterproof and insulated over-boots for longer trips.

Hat

Keep your head warm and dry with a reliable and comfortable hat. If you are sleeping out, carry two so you have a dry one for the night. Pack scarves or buffs (neck warmers) too; they can be used as hats if needed.

Snow play

Noses pressed to the cold window, eyes wide at the blanket of fresh snow out there … let's go and play!

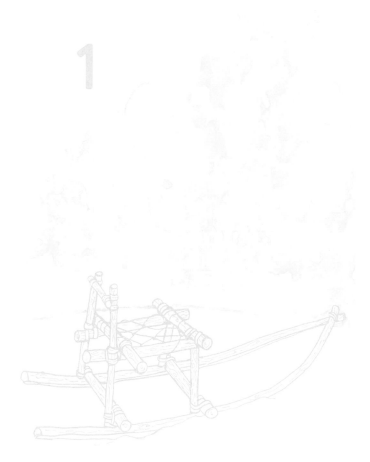

Make a sledge

It's snowing, grab the sledge ... or make your own! This is such an enjoyable process that involves creativity, tool work, trial and error. A lovely way to pass the day – and a sledge ride at the end of it.

1. First source some nice springy branches or trunks from spruce or pine (see page 80). Rather than using a sapling – a young tree – it's more ecologically conscious to find a recently fallen tree and harvest the branches while they are still flexible. Saw the branches to the rough lengths below (you can cut down to the exact size as you build).

Strip the smaller branches from these lengths. Smooth one side (the underside as they curve upwards) of the 2-metre runners with an axe.

2 branches 2 metres long and at least 6 cm wide at the thick end (runners)
6 branches 60 cm long and 3–5 cm thick (crosspieces)
2 branches 60 cm long and about 3 cm thick (back verticals)
2 branches 40 cm long and about 3 cm thick (front verticals)
1 branch 60 cm long and 1–2 cm thick (back rest)

2. Use a 2.5-cm auger or other drilling tool to make two holes in each of the 2-metre runners, drilling two-thirds of the way through the branch, at around 40 cm and 80 cm from the thicker end.

Drill a hole 6 cm from one end of two of the 60-cm crosspieces.

3. Whittle down one end of each of the 60-cm back verticals and both ends of the 40-cm front verticals. You want these to be just too big to fit into the holes you have made with the drilling tool.

4. Bash them in with a mallet or bigger stick until they are nice and snug, with the longer back verticals going in the holes 40 cm from the end of the runners and the shorter front verticals going in the holes 80 cm from the end of the runners. In the same way, attach the two drilled 60-cm crosspieces to the front verticals (figure 1).

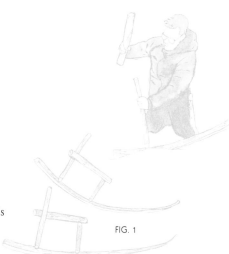

FIG. 1

5. Use square lashing to attach the remaining crosspieces and the back rest. When attaching the crosspieces on top, use your axe or knife to carve shallow V-shaped notches where they meet (figure 2). This will help the seat to sit securely and not wiggle. It is also useful to dig a small shallow trench under the sledge where you are lashing, so that you can get your hands and the string underneath.

FIG. 2

6. Finally, use netting to make a comfy seat. As an alternative, you could also use square lashing to attach further crosspieces to sit on. Now it's time to find some compacted snow and go for a ride … Have fun!

SQUARE LASHING

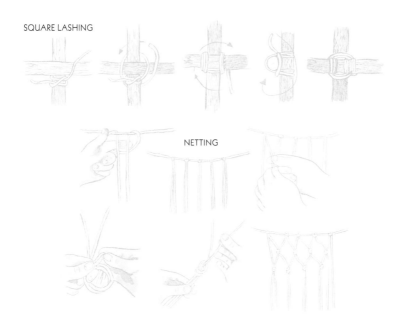

NETTING

Carrying children and gear

This sledge will also work well for transporting young children or gear – simply tie a rope to the front and get pulling (see page 20).

Other designs

I invite you to play with the design of this sledge and come up with your own. Here's some inspiration; the version below has no back rest, the seat is long and is tied to the runners at the front end and you can lie on your front while you ride. Can you think of other ways of improving it?

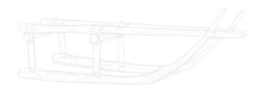

2

Treasure hunt

With its winter cloak on, the natural world can at first seem
homogeneous and uniform, but take a moment to let it all in
and you will see – there's treasure everywhere!

See

- Your breath
- Animal tracks (see page 104)
- Frozen puddles
- Berries on a branch
- Birds' nests
- Signs of spring, such as emerging flowers, blades of grass, buds on the trees
- Yellow snow (!)

Smell

- Pine needles
- Wood smoke
- Food cooking on the fire
- Cold air (how does it smell?)
- A hot drink

Hear

- Crunchy snow under your feet
- Dripping water
- The wind
- Birds singing
- Owls calling

Feel

- Snow falling onto your tongue
- Cold air in your lungs (take a deep breath!)
- Tree bark
- How heavy a piece of ice is
- Different textures of snow

Now, how many more treasures can you add to each heading?

Snow creatures

Everyone loves a traditional snowperson decorated with
coal and carrots, but why stop there? You can make animals,
fantastical beasts or even a huge snow creature.

1. Make snowballs in the rough shape of the animal.
They may need to be big!

2. Smooth out the joins of the snowballs and refine the shape,
adding more snow where needed.

3. Time to add the details; this could be stones for eyes and
noses, sticks for whiskers or a hat and scarf.

4. Why not add some treats for the birds? See page 62 for a
selection of bird-friendly treats that you could use to make
edible features on your snow creature.

4

Snow forts

It's a classic snowy pastime for a reason. Let's make a snow fort!

Please note these are for playing rather than camping out; for more practical snow shelters, turn to page 108.

1. Mark the outline of your snow fort on the ground using a stick.

2. Build the walls using mounds of snow, large snowballs or snow bricks (fill a plastic tub with snow, pack it down and tip it out to make a brick). Smooth the walls with a shovel or gloved hands.

3. What other features will you add? Turrets? Flags? A partial roof with sticks? This is also a good moment to douse water onto your fort so it freezes into a hard shell. You could also use natural dyes (see page 55) to give your fort a unique blast of colour.

4. Make your snowballs and store them in your fort. When you're ready, everyone into their forts and let the snowballs fly!

5

Snow maze

Age-old stories from across the globe speak of labyrinths as a
symbol of change and the personal journeys on which we are
each travelling. In the mazes of myth and the real labyrinths
that archaeologists have uncovered, we may enter, finding
our way before emerging changed …

This is a fun play on this idea, using snow to make our mazes.
It works well on a small scale with nothing but a pair of boots
on little feet, and can also be up-scaled to be as grand and
ambitious as you like. They are so much fun.

1. Find a flat area of untouched snow.

2. If the snow is deep, use a shovel, or in shallower snow, simply walk with your feet on the ground to clear a path into the snow. If you have enough time, energy and snow then you could build up your snow maze with taller walls.

3. Continue to build your maze, remembering to leave dead ends (paths that suddenly finish) to trick people …

4. When you are satisfied, invite people to come and walk your maze. Remind them to stay in the paths you have already made. Charge around in a wild frenzy or move mindfully through your maze while reflecting – it's up to you!

6

Kicker ramp

One of life's great pleasures. Build a ramp, then ride it!

1. Choose a good spot for your ramp. You'll need at least 10 metres of hill to get some speed up before you jump, and it's safer if you land on a steep (30–40 degrees) hill, too. You may be able to find a place with a natural ramp shape on which you can build.

2. Check the run-up and landing spot for rocks and sharp objects. Make sure your landing zone is smooth and long – you are going to fly further than you think!

3. Start to build your ramp at an angle of 20–25 degrees. It should be at least 1.5 metres wide and 2 metres long, although you can go as big as you like. A steep lip at the end of the ramp will launch you higher and a more gradual slope will send you flying further.

4. When you think you are done, pack it down and keep piling it on. You will need more snow than you imagine and want it to be nice and compact. If possible, let the ramp cool overnight to get nice and solid.

20–25 DEGREES

5. Get riding!

Play snow snake

This is a traditional winter game of many of the indigenous tribes of North America. It is simple and satisfying. A track is made by pulling a log through the snow to create a smooth, long trough, and then competitors take turns to see who can throw their snow snake the furthest along the trough. It's a lot of fun.

Traditionally, it would be played in four teams, with each player taking four turns and points awarded for the longest throw. A person's stick could be with them for decades and become a much-cherished object. On these pages, we look at how to make your very own snow snake stick. This one was made with Steve Le Say from Axe and Paddle in the UK, and we present this with respect and gratitude to the cultures to which this tradition belongs.

1. Find a length of straight-grained hardwood such as ash or oak. It can be anything between 60 cm and 2 metres long, depending on your size and preference.

2. Split the wood to roughly the size you want, remembering that your final stick should be wider than its height to prevent flipping. Ideally, you want a straight piece of wood but, if there is a natural bow, make this the underside of the snake. Draw the shape of your snake stick on the wood with pencil or charcoal.

3. Use an axe or knife to remove the excess wood until you reach the outline of your snake stick. Now, use an axe, knife, drawknife or, even better, the traditional mocotaugan ('crooked knife') to smooth and refine the shape. Your stick should have a smooth, flat belly, a rounded back and a head that looks like a snake's head.

4. At the opposite end to the snake's 'head', use your tool of choice to make a smooth indent on the end of the stick where a finger can sit for throwing (figure 1).

FIG. 1

When you are happy with its form, use sandpaper to take off any edges. Give the underside a final polish with a smooth pebble or burnishing bone; both tools will crush and harden the outer grain of the wood.

5. Now decorate your snake stick to make it uniquely yours: you can paint it with natural pigments, such as charcoal or ochre; play with making holes in it and inlaying heavy materials to help the stick fly down the course; or – as shown here – bash in some copper and brass tacks with a round-headed hammer to give the snake eyes and diagonal markings down its back.

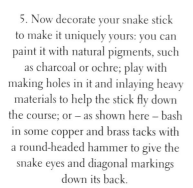

6. To finish your snake stick, make a 50:50 mixture of boiled linseed oil and white spirit and use a rag to rub the mixture all over the stick. Let it dry, then rub it with just the linseed oil. Let it dry again and it's ready to fly!

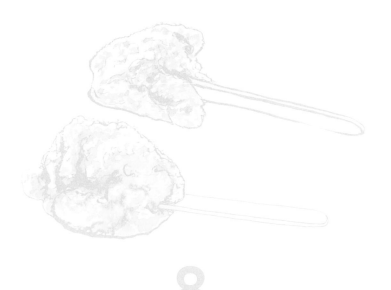

8

Snow treats

There is an old Jewish proverb that tells us, 'you can't make cheesecakes out of snow'. However, you can transform fresh powdery snow into delicious sweet treats with these fun recipes.

Only use freshly fallen white snow, stay away from ploughed snow and, of course, the yellow stuff!

If you have no fresh snow available or would prefer not to eat it, used shaved ice instead.

Snow ice cream

You are going to love this!

Makes about 1.7 litres
about 1.5 litres fresh snow
200 ml condensed milk
1 teaspoon vanilla extract or the seeds
from 1 vanilla pod

1. Mix all the ingredients together in a large mixing bowl until combined.

2. Get eating! You can freeze any leftovers, but it's so tasty, there might not be any left …

Snowflake tortilla

These snow-themed tortilla treats are fun to make and go well with a Mexican hot chocolate (see page 138).

Makes 1
1 tortilla
ground cinnamon, for sprinkling
sugar, for sprinkling

1. Fold the tortilla in half, then in half again, then in half a third time. Use kitchen scissors to cut out sections as if you were making a paper snowflake.

2. Open out the tortilla and sprinkle with ground cinnamon and sugar.

3. Cook on the fire in a frying pan with a lid covered in hot embers until the sugar is crisp. These can also be done at home in the oven.

Maple snow taffy

A popular snack in Canada, this recipe is simple and completely delicious. Children will need supervision and support as it involves boiling-hot syrup. You can spice up your taffy by adding ingredients like ginger, cinnamon, cayenne pepper or orange peel to your syrup as it heats. How about rolling your taffy in almond flakes or broken biscuits too?

Makes about 15
500 ml pure maple syrup
about 4 litres fresh snow
15 wooden lollipop sticks

1. Put the maple syrup into a pan and bring it to the boil over medium heat.

2. Very carefully pour the hot syrup in strips onto fresh clean snow (or snow in a baking tray).

3. Push a wooden lollipop stick into one end of each strip and twist it to collect the quickly hardening syrup.

4. Eat once cool and hardened.

Howl at the Wolf Moon

The full moon, which appears as a complete circle in the sky, sprinkles some magic onto any night panorama. The first full moon in January has come to be known as the Wolf Moon.

Opinions differ on the origin of this name. One of the more interesting theories is that this time of year marked the beginning of the medieval wolf-hunting season, when young cubs were vulnerable, and were targeted and prized for their pelts. It has also been suggested that wolves were driven from the forest into villages by hunger, and so more likely to be seen at this time.

Let's mark the Wolf Moon with a night walk, and remember the time when these enigmatic creatures walked the land. You may even be in a place where wolf populations are increasing or even thriving. If so, do take care and take precautions to keep yourself safe, and maybe you'll even have a magical wolf encounter!

1. Find out when the first full moon is in January.

2. Head out to welcome the Wolf Moon as the sun goes down. Pack a phone, torch and a flask of something hot. Notice how the landscape is transformed by the cloak of darkness and a sprinkle of starlight.

3. Sit and enjoy the calm of a bright moon in a cold winter sky. Take some time to reflect on your life, perhaps feeling gratitude for some of the good things you are blessed with. Get the flask out and have a hot drink.

4. Howl at the Wolf Moon! Awwwooooooooooooooo!

Snow craft

The snow and other treasures gifted to us by the winter offer so many opportunities for creativity and developing new skills. Let's explore some of my favourites …

Snow art

Snow is a magical material to work with. I can't help playing
with it, allowing delightful and surprising patterns and forms
to emerge.

The magic lies in the many textures and qualities of snow,
each of which lends itself to different types of pattern
and construction.

- Hard, crunchy top snow that is asking to be 'cut' into solid shapes and structures

- Mushy, creamy snow that creates lovely patterns, such as the spiral opposite

- Powdery, dry snow that can be packed together and left to sit for an hour or two to become solid

- Bottom layers of snow that can be compacted and compressed into a malleable material for structures

- Icy pellets that can be squeezed together or sprinkled

How could you use snow to create and decorate? What forms will emerge as you dig and sculpt? Take some time and enjoy getting to know this wonderful material in a different way.

11

Winter mandalas

Mandalas are beautiful geometric designs organised around a central point. These are important within a number of religions, particularly Hinduism and Buddhism, and mandala-like designs can be found throughout history and across the globe, such as in the Aztec Sun Stone in Mexico or the Great Pavement in Westminster Abbey, London.

I find the process of making mandalas meditative, inspiring and invigorating. As with the snow art (see page 46), treat this as an invitation rather than a strict set of instructions. The works on these pages were made with the brilliant artist James Brunt, although you do not need to be an artist to make these; this is about you enjoying the process of creation and letting something special emerge from the natural materials around you. I offer this with respect to the cultures for whom this is a sacred practice.

1. Gather natural objects – berries, cones, leaves, conifer needles and anything else. You may make the decision to take only things that are not part of a living thing. Take your time collecting; this is an important part of the process. You may be surprised by how many wonderful materials are out there when you start looking.

2. Start in the middle. You could begin with a star, cross, circle or anything you like!

3. Choose a material or pattern to place all the way around the centre point of your mandala, then build the design outwards by repeating this process with new materials or patterns. Your mandala will grow in size until you get to a point where it tells you that it's time to stop tinkering; it is finished. Try not to worry about the final product or perfectionism, just lose yourself in the process.

4. Experiment with other shapes and patterns, or little mandalas. There is no right or wrong, so get creating!

Snowball slingshot

Also known as a 'shepherd's slingshot' as it was used for millennia to protect sheep from predators, the slingshot is perhaps best known from the biblical story of David and Goliath. Although its roots are as a weapon, here I invite you to enjoy the calming weaving process and then to get out there and have some fun firing snowballs. You will be amazed at how far they fly.

I first heard this idea from Danish woodsman Johnny Juhl and this one was woven with Steve Le Say from Axe and Paddle in the UK.

You will need:
12 pieces of hemp twine about 3.5 metres long (we used four black and eight white for some decorative flair and because it can help keep the weaving nicely organised)
1 piece of hemp twine about 6 metres long
6 pieces of hemp twine about 30 cm long

1. Hold the 12 lengths of twine together and find the centre point, then move your fingers 5 cm along and fix them together at this point with a clothes peg or clamp.

2. Working on the longer end, split the 12 threads into two bunches of six. Twist each bunch to the left, moving the left bunch over the right every couple of twists. Repeat until you have made a loop big enough to fit three fingers, then tie with a spare piece of thread. You can now remove the peg or clamp and place the loop over something sturdy.

3. Separate the 24 strands into three bunches of eight, in this case two white bunches and one black (tip: when separating the threads, pull one strand at a time to prevent tangles). Plait them tightly until you have a 60 cm plait.

4. Using a 'weft and warp' weave, tie the 6-metre thread (the 'weft') onto the middle bunch then weave it back and forth, over and under the existing bunches (the 'warp'). After the first of these weaves, remove the string we added in step 2. If you have a netting needle (shown in step 7) it could come in handy here as we begin to weave the new longer thread. You could also make a simple weaving board by tying each bundle to a nail to keep them separate while you continue the weave, as shown here.

5. After each horizontal row, push your weave up to the point at which it splits, so it is spaced evenly. After around 2 cm, divide each bundle of eight threads in half, so you now have six groups of four. Continue weaving for another 2 cm, then divide them in half again, so you now have twelve groups of two. After another 2 cm, divide them into twenty-four individual threads. This will make our weave widen to accommodate the snowball.

6. It's time to make the central section of the slingshot pouch. Divide the threads into six bunches of four. Take one of the 30-cm lengths of thread and tie it to a bunch of four threads using an overhand knot (see page 61), then wrap the thread tightly round the bunch for 5 cm before finishing with a double overhand knot.

7. Now we can repeat the 'weft and warp' process, but this time in reverse; start with one thread, then two, then four and finally eight, at intervals of 2 cm.

8. Plait the three bunches of eight threads again into a 60-cm plait (as in step 3) and finish by tying an overhand knot. At this point you can head out to play, or soak the sling overnight with a tennis ball or stone tied into the pouch to give it shape.

How to use your sling

Never shoot snowballs (or rocks) at people or animals. Practise in an open space. Put your middle finger in the loop and the knotted end between your index finger and thumb and hold the pouch in your other hand. Swing a couple of times before releasing the knot and watch your snowball fly!

Snow painting

You get two great projects in one here, as you make your
own plant-based watercolour paints and then head out to
paint on the snow, your natural 'canvas'.

You can experiment with all kinds of natural ingredients, but here are some of my favourites.

- Beetroot
- Spinach
- Turmeric
- Berries

- Coffee
- Red cabbage
- Hibiscus flowers

1. Roughly chop your chosen ingredient and place in a pot. Add enough water so that you have one-third veg and two-thirds water in the pot. For soft, fragile berries and fruits, use two-thirds fruit and one-third water.

2. If you are using berries or soft fruits, skip this stage. Bring to the boil over medium heat and cook until it is the desired colour. This can be done on the fire or on the kitchen stove and should take up to 30 minutes. Let cool.

3. Blitz the veg or fruits and water in a food processor or blender (or mash with a fork) and get painting!

14

Frozen bubbles

This is magical. Make your own bubble mixture and watch
the bubbles turn into mesmerising spheres of ice before
your eyes. Any sub-zero (below 0°C) temperatures will work
here, but the colder it is, the faster the bubbles will freeze!
You will need washing-up liquid and sugar for this.

1. Mix 1 cup water, ¼ cup washing-up liquid and ½ cup sugar and stir gently until the sugar is dissolved.

2. Gently blow bubbles using a bubble wand or straw and watch as they land and freeze!

The science

What's happening here? The bubble is made up of three layers; the innermost of the three is water, which is surrounded by a soapy layer on both sides. The sugar helps slow down the water evaporation, which gives it time to freeze. It is the water freezing between the soapy film that gives us this wonderful and beguiling effect.

15

Winter mobiles

Harness the power of a freezing night to make lovely ice mobiles. This project can be done all year round using a freezer, and while this lacks the magic of leaving it out overnight to discover in the morning, it does mean that you can experiment with many seasonal materials as the wheel of the year turns through spring, summer and autumn.

1. Gather winter treasures such as leaves, berries, pine cones and other natural materials.

2. Place the materials in the sections of a bun or muffin tin. If you don't have one, you can use upturned jam jar lids laid in a row.

3. Take a length of string and run it through each section of the tin or jar lids, making sure that it goes fully into each section. Add water to cover.

4. Leave outside on a night when the temperature is below 0°C, or put in the freezer.

5. When they are frozen, carefully remove them from their moulds, then hang them up and admire your beautiful creation!

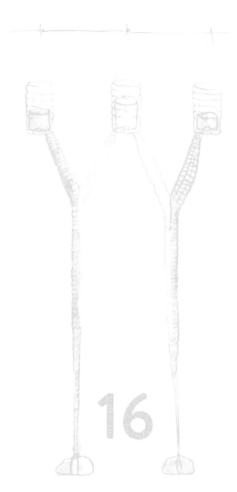

Coloured icicles

This is a cool winter science project that will create colourful icicles. We are creating a system where coloured water can drip slowly down a line as it freezes, with rich and vivid results on snowy days. You'll need sub-zero (below 0°C) temperatures for this one!

You will need:
3 large plastic bottles
2 lengths of thick string about 1 metre long
2 lengths of thinner string 1–2 metres long
Natural paint of your choice, shop-bought or homemade (see page 54)

1. Cut off the top half of each bottle and make a small hole in the bottom.

2. Thread one of the thick 1-metre pieces of string through two of the bottles before tying a simple overhand knot at each end to hold it in place. Repeat, connecting a third bottle with the second 1-metre length of thick string.

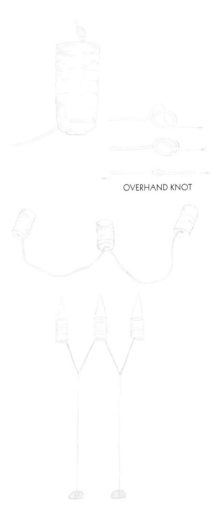

OVERHAND KNOT

3. Hang the bottles up (maybe on your porch or in your garden) and tie a thinner string in the centre of each of the thicker strings. Secure the string to the ground or to an object such as a stone.

4. Finally, add natural paint to the bottles. Leave it to drip and it will form amazing coloured icicles!

Bird feeders

Winter can be a tough time for our feathered friends, as the ground is too hard to dig for food and the shorter days afford less time for finding those all-important fats and nutrients. We can help them out in the colder months by making these all-natural nutritious treats.

You will need:

pine cones

100% peanut butter
(no sugar or salt) or lard

bird seed mix (or make your own with crushed peanuts, nyjer
seeds, sunflower seeds, cracked corn and dried fruit)

1. Coat the pine cones in the peanut
butter or lard.

2. Roll in the bird seed mix.

3. Hang your feeders from a tree and
sit back to watch the birds come and
enjoy the treat you have prepared!

Tip: You could also find an
identification sheet for local birds
and see which you can spot across
the seasons.

Snow lanterns

You will love these glowing snowball pyramids. Lit from within
by candlelight, they are beautiful, simple and so effective. Maria
Nilsson, the artist who created the wonderful illustrations in this
book, remembers making these snölyktor growing up in
her native Sweden. Try it!

1. Start with a lit candle. This could be as simple as a tealight on the snow, but may work better in a candle holder or jam jar to protect it from the wind. You could even use multiple candles.

2. Build a circle of snowballs around the candle, before building up the walls in a pyramid shape. It's a magical moment, putting that final snowball on top.

You may find that the structure wobbles or even collapses sometimes … it's OK, this is part of the process, so learn from it and go again!

Winter wisdom

The winter brings with it cold and darkness. It also brings
infinite gifts of knowledge and connection to the natural
world, as well as an opportunity for self-connection, rest
and growth. Will you resist and hide from its charms or
will you lean into it, surrendering to the call of winter?

Snow under the microscope

We can all conjure the classic image of a snowflake in our minds. But did you know there are a vast amount of different snowflakes? They are both surprising and utterly breathtaking in their diversity and beauty. Let's look at some of my favourites, and how you can spot them.

Which microscope?

A simple cheap magnifying glass of at least 5X magnification will work well for much snowflake hunting, and fits in your pocket.

A jeweller's loupe will cost more but will give you a clearer, more defined image. Go for at least a 7X rating.

The most expensive option, but one that will yield the most detail and impressive results, is a microscope giving you a field of vision of around 3 mm and at least 20X magnification across two eyepieces.

You can catch your snowflakes on a piece of cardboard or coloured packaging before having a closer look. See over the page for a guide to some of the extraordinary shapes you might find.

Diamond dust crystal

Also known as ice crystals, these form close to the ground in extremely cold weather and sparkle magically in the sunlight.

Stellar dendrite

Star-shaped snow crystal with six branches and lots of branches coming off each of those. Dendrite means tree-like; do these remind you of the phenomenon described on page 74?

Triangular crystal

These are wonderful and mysterious – science has still not entirely worked out what conditions cause the formation of these cool crystals.

12-sided snowflake

These occur when two six-sided snowflakes collide in midair and stick together!

Capped column

These wonderful shapes form when the snowflake travels through different temperatures as it grows.

Rimed snowflake

These are snowflakes that have collided with tiny water droplets, called rime.

20

Finding nature's numbers

Out in the snowy winter, the magic of nature's repeating patterns is all around. Let's look a bit closer …

The Fibonacci sequence

This fascinating sequence was presented in 1202 by Leonardo Fibonacci and is very simple: start with number 1 and add the previous two numbers (1, 1, 2, 3, 5, 8, 13, 21, 34 and so on). We can also see the Fibonacci spiral emerge as we draw quarter-circle arcs across the numbers from his famous sequence. The number sequence and spiral crop up everywhere in nature, from the number of petals on flowers to shells, cobwebs and even galaxies!

In the snowy winter landscape, find a pine cone, look at it from the bottom and count the spirals going in each direction. You will find that the number of spirals going in both directions can be found in Fibonacci's sequence.

Where else can you find these magical numbers in nature?

Tree fractals

Fractals are geometric shapes that repeat their patterns in ever-finer scales. Think of a river. Its natural branching form, splitting off into smaller and smaller streams, is a fractal pattern just like the shape of a cauliflower or the branches of a tree.

A deciduous tree (see page 78), when viewed in the winter without leaves, reveals its glorious fractal architecture. The pattern is simple: the trunk splits into two branches, which then split into two smaller ones and so on. Fascinatingly, most trees have no more than 11 of these splits on any one branch. Head out and count them!

Da Vinci's Rule for Branches

Da Vinci wrote, 'All the branches of a tree at every stage of its height when put together are equal in thickness to the trunk.' Put another way, if you cut the branches at any one height on the tree and added them together, they would add up to the same size as the tree's trunk. This is mind-blowing – it means that if the tree were squeezed upward with all the branches brought into the centre, it would be the same thickness throughout! It has since been shown not to be true in absolutely every case, but is still very cool!

FROM DA VINCI'S
NOTEBOOK,
15TH CENTURY

Rise above the clouds

To stand above the clouds on a mountain top is to inhabit
another world. It is a truly breathtaking sight.

Imagine hiking up through the fog and emerging on foot to find
a fluffy white sea of cloud at eye level or unzipping your tent in
the morning to discover this magical phenomenon. Or bursting
through the clouds on a ski lift before snowboarding back down
through them. However you do it, spending time above the
clouds is magical and unforgettable.

Cloud inversions, more common in the colder months as the
nights are longer and the air cooler, are caused by a combination
of meteorological conditions: temperature inversion (a reversal of
the usual pattern of temperatures falling as you get higher), high
moisture levels, calm conditions and high pressure. This means
the clouds sit below the mountain tops and we are able to rise
above them, into an otherworldly landscape.

So the next time you look out of the window and wonder
whether to head up the hill on that grey day, just remember:
a cloud inversion could be waiting for you up there.

Ice bathing

Ice bathing (and, more generally, immersion in cold water) has been rather in vogue in recent years thanks to high-profile proponents of its many benefits such as Wim Hof.

The custom of exposure to the cold for health and wellbeing is, of course, much older than this; for example, Scandinavians let their babies nap outside in low temperatures to improve sleep and boost immunity, while the cold can be medically prescribed for wellbeing in Russia.

Cold-water bathing has been scientifically proven to speed up your metabolism, improve quality of sleep and strengthen the immune system. You may also experience a profound sense of calm, gratitude and vitality when leaving the cold water (even if you don't always feel like getting in before your dip!).

Nordic and Scandinavian winter swimming clubs traditionally cut holes in the ice so that people can enjoy the endorphin kick of hopping straight from the sauna into the freezing water, or even rolling in the snow. Bliss! You don't need a club to try ice bathing, but do take note of these practical tips:

- Start with a quick plunge and work up to staying longer in the cold water.

- Keep your head warm with a hat and don't submerge your head if you are a beginner.

- Don't go alone.

- Consult a doctor before taking the plunge if you have a history of high blood pressure, heart problems or diabetes.

23

Winter tree identification

As we wander the snowy landscape and see the trees standing bare and naked without their leaves, it is easy to think of the winter as a time of death … but this is not the case!

Deciduous trees – those that lose their leaves for part of the year – are very much alive, using the colder months to rest, restore and strengthen. It is a practice that we humans can learn a lot from, and is discussed further in the exploration of wintering (see page 90).

The buds of deciduous trees, from which new leaf growth will emerge in the spring, are a wonderful illustration of the life in winter trees just waiting to emerge into the sun. These buds are, in fact, already on most trees in the autumn as the old leaves fall, protected from the cold by thick scales.

These tiny wrapped parcels of potential, containing minuscule leaves and flowers, are also great ways to identify deciduous trees in the winter; take this book out with you and see which you can find.

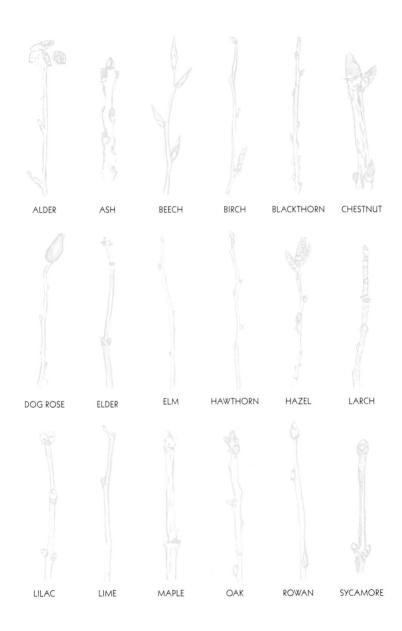

ALDER   ASH   BEECH   BIRCH   BLACKTHORN   CHESTNUT

DOG ROSE   ELDER   ELM   HAWTHORN   HAZEL   LARCH

LILAC   LIME   MAPLE   OAK   ROWAN   SYCAMORE

24

Conifer identification

Picture a snowy winter wonderland scene in your head and it's a fair bet that there are conifer trees dotted around the fantasy white landscape.

Characterised by their thin, waxy needles and cones, conifers are often the only trees with a green canopy in the winter, though not always (both larch and dawn redwood drop their leaves in autumn).

Conifers thrive in the snow. This is partly due to their drooping limbs, which can support a heavy snow load, but also thanks to a number of anti-freeze measures, such as needles full of a concentrated sugary sap and thin tracheids (the tubes that carry water around the tree).

CORSICAN PINE

DOUGLAS FIR

EUROPEAN LARCH

GRAND FIR

JUNIPER

LAWSON'S CYPRESS

LODGEPOLE PINE

Pine, fir or spruce?

Though not the only types of coniferous tree, these three varieties are very common and it's useful to be able to differentiate between them.

We can tell the difference between pine, spruce and fir by their needles. On a pine, the needles are clustered on the branch in sets of two, three and five (here is Fibonacci's sequence again, see page 73), whereas fir and spruce needles are attached individually to the branch.

To tell the difference between a fir and spruce, roll the needle between your fingers – spruce needles are four-sided, so they roll more easily than the flat needles of a fir.

The needles of some conifers can be used to make a delicious tea, so get the kettle on and turn to page 136 to learn more.

NOBLE FIR

NORWAY SPRUCE

SCOT'S PINE

SITKA SPRUCE

WESTERN HEMLOCK

WESTERN RED CEDAR

YEW

25

Spot crown shyness

As your boots crack through a frozen puddle and crunch
through the snow on a winter walk in the woods, take a moment
to look up and spot one of nature's many wonders – crown shyness.

This occurs in a forest when the crowns of trees avoid touching
each other, creating boundaries and clear channel-like gaps
in the canopy. There are a number of different theories as to
the cause of crown shyness, such as trees minimising harmful
competition for light or inhibiting the spread of leaf-eating
insect larvae and disease. Whatever the reason, the results
are quite beautiful. Enjoy.

See the northern lights

The northern lights (*aurora borealis*) are an atmospheric phenomenon that has captivated us for millennia, an ethereal light show dancing across the winter sky.

What causes the northern lights?

It's fascinating. Charged particles emitted by the sun are carried toward us on the solar wind. These particles interact with the magnetic field around the earth and are pushed toward the polar regions. When these charged particles collide with our atmosphere (at a speed of 45 million miles per hour!) they release energy in the form of light.

The colour of the northern lights is determined by the element with which the particles collide. Each element in our atmosphere will give an entirely unique colour, oxygen giving a greenish glow and nitrogen a reddish hue. We see a lot of greens in the northern lights due to the high levels of oxygen in the upper atmosphere.

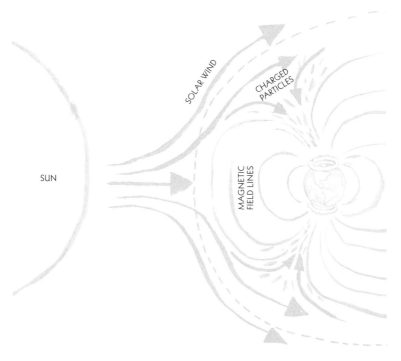

When to go

The northern lights occur all year round, but are best viewed from November to March when it is darker. In the southern hemisphere, you can see the southern lights (*aurora australis*) between May and August.

Where to go

Head to Iceland, Norway, Sweden, Finland, Russia, Alaska or north-western Canada and find a spot away from light pollution in an open space. As we learned above, this phenomenon occurs at both poles, so you can also catch the southern lights in Australia, Tasmania, New Zealand, Patagonia and Antarctica.

Further afield, auroras also occur on other planets! So if you are wintering on Jupiter, Saturn, Uranus or Neptune, do be sure to check them out …

Find the Winter Circle

It seems that humans have always stared at the night sky.
Cultures throughout history recorded their own versions of
the luminous dot-to-dot shapes gifted to us by the starry
panorama, from bear to ram.

Winter brings longer nights, and you may find your attention
turning skyward to those twinkling balls of energy and mass,
light years away. As with our ancestors, there seems to be a
tendency in us to find order in the dark ceiling of the night sky.
I offer you one such collection of stars that is delightfully easy to
spot in the colder months – the Winter Circle, formed by six of
the brightest stars in the northern hemisphere's winter sky.

Spotting the Winter Circle

Start by finding the easily recognisable set of three stars that make up Orion's Belt. Look lower and to the right, and you'll find Rigel, the first of our six winter stars. If you follow the straight line of Orion's Belt you'll find Aldebaran, the Eye of Taurus from the bull-shaped constellation. Moving anticlockwise, we find Capella, the sixth brightest star in the sky, before continuing round our circle to Pollux, the brighter of the Gemini 'twins'. Finally we meet Procyon, before ending gloriously with Sirius, the brightest star in the entire night sky. Its name means 'glowing' in Greek, which is well earned as only the full moon, a few planets and the International Space Station outshine this star!

The Winter Triangle

Let's also give a mention to the Winter Triangle, a pleasingly equilateral triangle drawn on the celestial sphere between Procyon, Sirius and (in the centre of the Winter Circle) the awesome Betelgeuse, not just a star but a red supergiant, 1.4 billion kilometres across and emitting as much light as at least 50,000 suns!

28

Winter forest bathing

Forest bathing, based on Japanese Shinrin-yoku, involves
spending time among trees to improve wellbeing, reduce stress,
strengthen your immune system and lower blood pressure, as
well as deepening your connection to the natural world. Winter
offers a magical time to enjoy forest bathing with the sparkle of
snowfall and exposed tree architecture.

Rather than instructions, these are invitations to go out and be
present in the moment, in your body and in that place; like a
more mindful and conscious version of the sensory treasure
hunt on page 26.

See

Start by focussing on a point in the distance – a snowy peak or faraway tree – before slowly bringing your vision closer to where you sit or stand. Trace the shape of a tree with your eyes. Notice tree fractals and Fibonacci spirals (see page 72). What is moving and what is still? End by focussing your attention on the tiniest things right next to you, perhaps even a single snowflake (see page 68).

Hear

We call this one 'deer ears', as we mimic a deer's large ears and their wonderful capacity to change the direction of their hearing. Cup your hands behind your ears and let in the sounds of the winter forest. Move your hands around, getting a 360-degree aural immersion. Running water can seem louder when our 'deer ears' are cupped to listen behind us – so put your back to that semi-frozen stream and marvel at this strange effect!

Smell

Go on, invite in your sense of smell – dig some earth and let its aromas fill your nostrils. Smell sticks, leaves and the wood smoke from the fire. Winter is a rich olfactory landscape when you lock into your sense of smell.

Touch

Take your gloves off for a while and marvel in the many textures of the woods in winter. Explore tree bark, run your fingers along a conifer branch, snap dead sticks or wrap your arms around a tree and give it a hug!

Taste

So much of our eating and drinking is done on autopilot. Focus truly on the sensory experience of sipping a hot tea. You may experience an everyday act as if for the first time. Try it!

Practical considerations

Central to forest bathing is slowing down, both body and mind. This stillness, together with cold winter conditions, means that you must protect your body temperature, so wear warm clothing (see pages 9–19) and take a hot drink.

29

Wintering

It can be tempting to fight against the winter and wish away the colder months when the snow falls and the sun hides behind clouds. But what would happen if we leaned into the winter and its gifts rather than wishing it were summer? What would happen if we did as nature does and used it as an invitation to rest and restore, adapt and prepare?

Rest and restore

Do as the trees do (see page 78) and take time to ground,
strengthen and spread your roots. Take time to sit with your
soul – tuning into your inner world, as well as the natural world
outside your window. Watch winter for its messages – slow down
and let your spare time expand as you get cosy and hibernate
like a bat, hedgehog or bear. Give time to practices that support
your inner growth, which will be different for each of us, but
could include walking, reading, poetry, journalling, collage,
doodling, knitting, making mandalas (see page 48) or whatever
emerges from you. You deserve this time.

Adapt and prepare

As you open yourself to winter's lessons and its magic, ask
yourself, what do I need to thrive in winter? After all, winter
is a time of metamorphosis, not death. This is taught to us
beautifully by the snow fox, snow hare and ptarmigan, who,
every year, swap their grey-brown feathers or fur for a cloak of
brilliant winter white. What part of you is asking to change and
evolve? Looking ahead, how would you like to re-emerge into
the burgeoning spring after the deep retreat of winter?

Winter solstice

The shortest day and the longest night, the winter solstice usually falls on 21 or 22 December in the northern hemisphere. It marks an important moment for cultures across the globe and, indeed, it has across millennia. Let us focus on the Celtic tradition, native to Ireland, the UK and other parts of Europe.

The Celtic wheel of the year is a pre-Christian annual cycle of festivals celebrating the seasons and solar events (solstices and equinoxes). The winter solstice, or Yule, arrives in midwinter and marks the sun's disappearance from the sky.

It is a time to celebrate both the light and dark; the inner light in each of us that never goes out and gets us through the darkest and coldest of times, as well as the powerful and restful state of darkness itself, with its seductive stillness and secrets.

I invite you to honour the solstice with some ritual on the themes of darkness and light, using the ideas below, if you like.

Darkness

The pulse of the earth has slowed and the calm of the deep darkness is inviting us. You could camp out on the longest night, getting away from electric light and into the night. Come to know the darkness as an ally and a gift, appreciate its qualities and open your senses to its magic.

Light

Sometimes in midwinter we doubt the light will ever return, but it always does. Let's honour this by lighting a candle or a fire. Tell stories of loss and birth at this time of change. Ask yourself what needs to be shed and left behind, and what parts of you could be nourished and grown. Vocalise something that needs to be said, examined or celebrated, bringing it from the darkness into the light. Watch the sun break free over the horizon in the east. As is said at this time, 'we have turned the year!'.

Samhain and Imbolc

We can also honour the two cross-quarter point festivals that sit either side of the winter solstice – Samhain and Imbolc. Samhain comes at the end of October or beginning of November, and it invites us to begin our period of darkness and rest as explored in wintering (see page 90). Imbolc, which comes at the end of January or beginning of February, celebrates the return of the sun and the reawakening of the earth, as well as our own re-emergence from the snowy winter into the light of spring.

Winter survival

Like some of the approaches in the previous chapter, practising
survival techniques is an invitation to work *with* the snow and the
winter, rather than against it. While there may be connotations
of macho posturing and the like, these skills – apart from being
truly lifesaving – are also a route into a deeper connection with
the land and the ways of our ancestors. Being out in the wild and
unplugged from technology can make you feel truly alive.

Survival requires a slow and steady pace, eschewing panic in
favour of a methodical calm. It asks you to be present, to stay
connected to your body and surroundings. This is not a survival
manual – many such books exist – more another route into deep
nature connection that could, one day, save your life.

Rather than beating or overcoming nature, I urge you to think
of winter survival as a way to live more in harmony with nature,
even (or especially) in her harshest and coldest moments.

Make snowshoes

All winter survival techniques are best practised when you do not need them, so they may come in handy when you do. These snowshoes are a great example – I remember one passer-by in the woods looking at me with a surprised expression when I told him I was making snowshoes … in the height of the summer in London!

1. Find ten straight lengths of wood that are around as long as the wearer's height. Birch or spruce saplings will work best, but in an emergency situation, of course, use whatever you have to hand.

2. Carve notches into each length 10–15 cm from each end of all ten lengths of wood. Fasten five of them together at the thinner ends with jam knots (see next page) over the notches. You may need to use a knife to thin out and taper the ends. This is the front of the shoe. Repeat with the other five for the second shoe.

DIAGONAL LASHING

3. Find the centre of gravity of the rods by balancing them on your hand. Attach a crosspiece around 15 cm from the centre of gravity toward the thinner end using diagonal lashing. Rest your heel on this and mark where the ball of your foot sits. Attach a second crosspiece at this point using diagonal lashing again. Repeat for the second shoe.

JAM KNOTS

DIAGONAL LASHING

4. Use jam knots to fasten the snowshoes at the back, in the same way as you did at the front. Repeat for the second shoe.

5. Attach a length of paracord or other cordage from the front crosspiece to the front ends of the snowshoes to bend them up at the front. When they have dried out over the following hours and days, you can remove the string and the bend will remain.

6. Finally, create your bindings with a length of fabric or paracord as shown here (figure 1).

FIG. 1

32

Winter navigation

Knowing the basics of navigation is helpful on a mild summer's day and can be the difference between life and death out on the winter trail.

In the snow, paths and trail signs are obscured or buried, so knowing how to use a map and compass is essential. As well as hiding a known route from us, the snow also gifts us the possibility of traversing terrain that is otherwise not an option, such as frozen lakes (see page 113) or marshland.

Carrying a GPS is a good idea if you can, as this can tell you exactly where you are in an instant. However, as these can malfunction or run out of battery, we will focus on the classics – knowledge, a map and a compass.

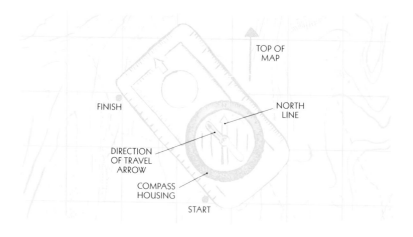

TOP OF
MAP

FINISH

NORTH
LINE

DIRECTION
OF TRAVEL
ARROW

COMPASS
HOUSING

START

Following a bearing using a map and compass

1. Find the points you want to travel to and from on the map, and place the compass flat on the map so it forms a line between the two, making sure the direction of travel arrow is pointing in the direction you plan to travel.

2. Rotate the round compass housing so the north line is aligned with the vertical north–south grid lines on the map.

3. Line up the compass needle with the north line.

4. Walk using the direction of travel line toward your destination.

If visibility is good, pick a landmark that is in your direction of travel, such as a large tree or boulder, and use that to stay on track, stopping to take new bearings regularly. Even better, use the 'three object sighting' method, explained opposite.

In poor visibility such as a whiteout, walk as a group in single file and keep checking behind you – if the line is curved, you are going off course.

Three object sighting

This is a simple method of travelling in a straight line by choosing three objects in a row in your direction of travel. Walk towards the middle object, checking that you have all three in line. When you reach the middle object, turn back to check you are still on course. Now choose a new third sight point in the distance and continue in this way until you reach your destination.

Pace counting

For shorter distances and especially in poor visibility, you can use pacing to estimate how far you have travelled. Work out on flat ground how many steps it takes you to travel 100 metres, then use that as a guide as you move. This will change depending on the terrain, but with practice can be an important tool.

33

Staying found

Let's delve further into the topic of snow navigation, explored
on the previous pages. At this point, I invite you to try two
approaches that will deepen your understanding of your
environment and sharpen your navigational senses. These are
more about how to 'stay found' rather than what to do if you
lose your way, but they could stop you getting into a survival
situation at all. They can be practised all year round. The first of
these may seem counterintuitive at first … it's time to get lost!

Get lost

Pack a bag with everything you need (see pages 9–19) and set off with at least two other people. You can all take turns getting lost on purpose and finding your way back to a chosen spot while the others stay within earshot, so you can shout if you really need help. Getting lost will help you familiarise yourself with the feeling of uncertainty in your stomach, and you'll be less daunted should the real thing occur. It also might make clear that there is a skill you need to practise or a piece of equipment to make or buy. By improving your navigational skills and awareness of your surroundings, you'll increase your ability to navigate your way out of being lost, and to stay found.

Story trails

This is a technique practised in different forms by people across the globe and is a beautiful way to stay found, by being truly attentive to your environment. Invent a story when walking into an unknown wild place using its natural features – rock formations, clusters of trees and so on – to inspire the events in your tale. Then, at some point, turn around and use the story to 'tell' your way back in reverse. I love this one.

Snow tracking

Snow is a wonderful surface for finding animal tracks. A light dusting of snow can highlight even the subtlest paths and deeper snow retains a pristine print. In a survival situation, this can lead us to water (see page 116) or a food source through trapping, hunting or foraging.

We can also enjoy finding and following these prints simply for the joy of it. This practice of tuning into nature's signs is an ancient one that is both entirely natural to us and a skill that takes time to develop. Ask yourself: Who left this track? When was it left? What were they doing? Why did they pass through here? Where were they going? Even, how did they *feel*?

On this page is a selection of animal and bird prints that you may find in the snow, depending on where you are. Of course, animals and humans leave many 'prints' beyond footprints, such as nesting sites, scat, pellets, food debris and more. Open your eyes and senses to these things and, in doing so, you can connect more deeply with the land and the animals within it.

BADGER   BEAR   BEAVER   CROW   DOG

DUCK   EAGLE   FERRET   FOX   GOAT

HERON   HORSE   HUMAN   LYNX   MARTEN

MOOSE   MOUSE   OTTER   OWL   RACCOON

SQUIRREL   RED DEER   ROE DEER   WOLF

35

Winter foraging

Winter can seem a barren time
when it comes to things growing,
but there is a surprising amount of
sustenance to be found out there if
you know where to look – gifting us
countless treasures. Here are just a
few of them. Remember, forage from
unpolluted areas and don't eat plants
unless you are 100% certain: 99%
sure is not enough!

Bulrush tubers

Found around water and
distinguished by their fluffy sausage-
like flower heads, bulrushes are easy
to spot. Dig up their roots and peel
them, then they can be boiled or
fried. However, do be careful not
to get water into your boots when
collecting these … that's no fun
on a cold day!

Jerusalem artichokes

Growing wild in fields and on forest edges, you can spot Jerusalem artichokes by their spiky dried flower heads that stand on tall stems. Dig up and clean the tubers before enjoying them raw or cooked using your preferred method.

Rosehips

Roses are a brilliant source of winter nutrition. You can identify rose plants by their thorns and bright red rosehips, which are high in vitamin C and antioxidants. Squeeze them between your thumb and forefinger to enjoy the flesh, but don't eat the skin or seeds. Eat raw or brewed up to make tea, jam, syrup or even wine!

Evergreen teas

Turn to page 136 for a delicious recipe for evergreen tea.

36

Snow shelters

Making snow shelters is a huge
amount of fun if you have the time,
energy … and snow.

Snow cave

1. Choose a spot high up the snowdrift,
being careful to avoid high-risk areas
for avalanches. Use a shovel or saw to
dig out a small entrance into which
you can easily crawl and fit your gear.

2. Excavate a domed area with
a raised bed on which you can
comfortably sit up. This will create
a 'cold sink' to collect the cooler air
and keep you warm. Your walls
should be at least 20 cm thick.

3. Add an air vent at a 45-degree angle (remember to clear it at intervals with your ski or trekking pole) and a shelf for a candle. This will provide light and, if it flickers or goes out, an important warning that oxygen is low in the shelter.

4. Keep the roof smooth and curved to prevent drips and place a snow block, cut from ice with a saw (page 10), in the entrance. Remember to sleep with your shovel inside the shelter, and mark the area above your cave, as you don't want to walk on the roof!

Snow trench

A snow trench is quick to dig and a cosy spot to spend the night. In its simplest form, a snow trench is a ditch 1.5–2.5 metres deep and around 70 cm wide. You may wish to shape steps into your trench. Finally, cover with excavated ice blocks, branches from nearby trees or a tarpaulin.

Shovel substitutes

In the absence of a shovel, you could use a snowshoe or a stick. People across the world have used roughly sharpened sticks for excavating for millennia. Or how about finding a strong Y-shaped stick then weaving smaller sticks into it? Or you could hang a T-shirt on that stick and dip it in a water source before letting it freeze. And if the water supply is low, you could even urinate on it …

Turn the page for a look at two magnificent and ancient snow shelters.

Igloos and quinzees

Let's take a look at two classic dome-roofed snow shelters:
the igloo and quinzee.

The igloo is an iconic design, traditionally used by the people
of the Canadian Central Arctic and Greenland. Simpler in its
construction, but similar in shape to the igloo, the quinzee has
been used by indigenous peoples across snowy North America
and Canada for many a year.

An igloo will work well using icy, packed snow that can be sawn
into blocks, whereas a quinzee can be useful in the forest where
the lighter snow can be piled and compacted.

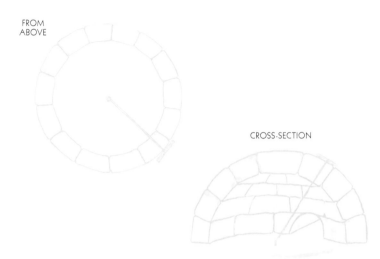

FROM
ABOVE

CROSS-SECTION

Igloo

1. Saw ice blocks around 80 × 40 × 25 cm in size.
You may need up to 40 of them.

2. Drive a stake into the ground with a piece of string
attached. This will be the centre of your igloo and the cord
will help to dictate proper angles as you build.

3. Place your first block and mark a point or tie a knot or stick
onto your string where it meets the outside of this first block.
Continue to build the igloo, making sure the outside of each
subsequent block meets this point as your structure grows.
Use your saw to cut the blocks at an angle in line with the
string too. The person building the igloo stays in the centre,
while others saw and collect ice blocks.

Quinzee

1. Make a pile of powdered snow 2–3 metres wide and 2 metres tall, packing the top 30 cm of the dome more tightly with a shovel.

2. Collect sticks 30 cm in length and poke them into your pile at regular intervals.

3. Excavate a large entrance at the base, before digging out the snow inside the dome until you reach the sticks.

4. Finally, make the doorway smaller with sticks or snow and get comfy!

Crossing ice

It can be tempting to cross a frozen body of water rather than walk the long way around it. Lakes and ice are often crossed on winter journeys and some routes across lakes are so safe that even cars drive across them. However, if it is not known to be safe, it's wise to only do so if there is no safer alternative. Never allow children to play on ice unless the authorities have confirmed its safety. Let's take a look at the precautions you can take, as well as what to do if you fall through the ice.

Precautions

- Check local notice boards for information about ice conditions before heading out.

- Check the thickness of the ice in several locations by using an ice auger or ice checking pole/spud bar. It is never safe to walk on ice that is less than 10 cm thick (this applies to clear ice only).

- Avoid areas where you can see cracks, holes or water moving below/near the edges of the ice.

- Don't approach open water.

- Clear ice is stronger than cloudy ice.

- Travel in a group, with the heaviest person at the front.

- Take small steps.

- Carry 15 metres of rope or a throw line and ice axe (see pages 13 and 12), plus an ice pick worn around the neck.

- Keep a dry change of clothes in your rucksack in a sealed plastic bag; the trapped air will keep you afloat and you'll have a fresh set to change into if wet.

- If you see or hear cracking, lie down and spread your weight.

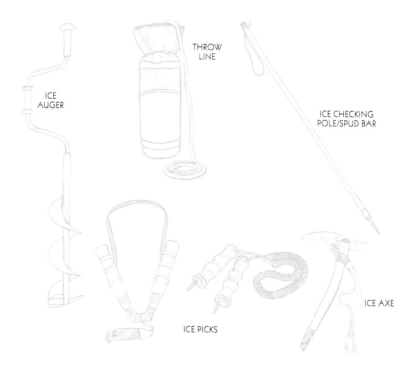

ICE
AUGER

THROW
LINE

ICE CHECKING
POLE/SPUD BAR

ICE PICKS

ICE AXE

If you fall through the ice

1. Stay calm and get moving; every second counts. Turn in the direction you came from, as you know the ice can bear your weight there.

2. Put your trekking poles under your arms and use your ice axe, ice picks, knife or pole to get a firm grip on the ice.

3. Kick hard in the water as you pull yourself up with one knee up onto the ice.

4. Spread your arms and legs and crawl back to safety. Don't get up until you know you are on firm ground or safe ice.

5. If possible, other group members should make a fire on the shore and lend dry clothing to the person who went in. After a warm drink they should begin moving around to raise the body temperature.

Finding water

As the Spanish proverb goes, 'from snow, whether baked or boiled, you will get nothing but water'. Although not strictly true, as evidenced by the delicious treats on page 39, snow and the winter landscape provide a number of important water sources that can save your life and certainly prevent you carrying heavy supplies.

Snow melt

The first time you see how much snow it takes to make a small amount of water is often surprising, and illustrates the fact that snow is actually full of air. Use fresh white snow, melting a small amount first then adding more. Boil for at least 3 minutes.

Natural springs

Find a spring by moving upstream from a small creek or tributary, which may make its presence known by the sound of flowing water or a place where snow is not settling. You may even find a spot where fresh water is literally coming out of the ground. Consider whether the source could be polluted or contaminated and, if possible, boil your water for at least 15 minutes before drinking or use a water filter.

Tip: Dip your bottle or container into water facing away from the direction of flow and you should get less sediment in your water.

Tapping trees

Another great source of hydration is 'tapping' trees, such as birch, maple, walnut and sycamores, to extract the delicious and sugary sap in late winter and early spring. While this can be done by drilling or cutting into the tree trunk, this is an intrusive method that can harm the tree. I prefer to find a low branch and make a diagonal cut before hanging a container to catch the sugary sap.

Snow fires

Fire and snow – what a combination. Lifesaving, life-affirming and one of life's great pleasures. In this section, we explore the many possibilities for fire out in the winter wilds.

40

Finding firewood

In the deep snow, it can feel as if everything is soaked and frozen. But there's always dry wood for your fire if you know where to look.

Dry wood

For bigger pieces of wood, you are looking for clues that it is dry and dead, such as a greying colour and long, deep cracks along the branch or trunk. It will also give a satisfying sound like a cricket or baseball bat when struck. Smaller twigs should snap when bent. Wood off the ground or the dead lower branches from live conifers works well. You could take down a whole standing dead tree if you need that much wood. Remember too that inside most wet dead wood is dry wood, so split it with your knife or axe to strip away the soaked outer layers.

Fatwood

Most coniferous trees (see page 80) contain flammable resin. When a side branch dies, this resin will collect at the point at which the branch meets the tree, as well as the base of the trunk. Split the wood in these areas with your axe and knife. It should smell like turpentine, with darker veins inside. Create shavings with the cutting edge or back of your knife, then light them with a fire steel before feeding and growing your fire.

USNEA
LICHENS

Tinder

Tinder is the highly combustible material we use to start our fire, before burning kindling and then progressively larger bits. Useful tinder includes shaved fatwood (as above) or birch bark, dried seed heads, usnea lichens and fungus, such as cramp balls. Keep an eye out for these as you travel, then collect and store them in your tinder pouch (see page 13).

Wet wood

If you have to use wet wood for your fire, hold bundles of thin kindling (small sticks up to the width of a pencil) above the flame to dry them out before adding them to the fire. Then reach for another bundle of thicker sticks and do the same. When your fire is roaring, you'll be able to add wet and even frozen wood, though they will burn slowly and produce more smoke.

Making fire

You've gathered your materials (see page 120) and you've got your fire-lighting kit (see page 13), so let's get that fire going!

Siting your fire

Choose a well-drained, sheltered spot close to your fuel source with no snow-laden branches or hanging dead wood overhead.

Dig!

Unless you are just having a quick fire, you'll want to dig down, preferably to the ground. This will provide shelter for you and your fire and a reflective wall for extra heat, as well as avoiding the fire sinking directly on the snow. Dig out a bench that can comfortably sit each member of the group.

Upside-down fire

This is a great way to build a fire in the winter, as it can be done directly onto snow or the frozen ground and, after lighting, you shouldn't need to tend it for a while. It works in the opposite way to the standard method of making a small fire then adding the larger bits. Instead, we build a structure with the biggest logs at the bottom before building upwards with progressively smaller fuel wood and the smallest twigs and tinder on top. The fire then burns down in a controlled and long-lasting blaze.

Leave no trace

Unless it is a permanent fire site, make sure you scatter all evidence of your fire and fill in any pits you have dug before you leave.

Char cloth

A very useful resource for fire lighting, a charred cloth will catch
a spark easily, providing a glowing ember that can be placed in a
tinder bundle and blown into flames. It is made by burning a piece
of 100% cotton fabric (or fabric made from other plant fibres) in
an oxygen-deprived environment – a simple and satisfying process.

1. Find a tin with an airtight lid. Make a hole by resting a screwdriver or tent peg in the centre of the lid and striking it with a mallet. Cut your cloth into pieces 20 cm square.

2. Put your cloth into the tin in a loose bundle then firmly close the lid and place in the fire. Be careful not to breathe in the resulting smoke.

3. When the smoke stops, carefully remove the hot tin from the fire and plug the hole with a stick. If your char cloth is not fully black, simply put the lid back on and place the tin back in the fire until it is fully charred.

4. Store your char cloth in an airtight container until you're ready to use it.

Ice fires

At first there seems to be some kind of elemental magic at play here, as we use ice to make fire. In fact, this is a simple variation on the classic solar fire-starting method, where the sun's rays are concentrated and redirected to create a fire. It takes some practice, but you'll get there in the end. Have a go!

1. Finding clear ice (without bubbles) is an important first step. The ice around a river or lake could work well, or you could even make your own by boiling water, leaving it to cool twice before freezing.

2. Shape the ice into a convex lens form with a knife and smooth with the heat of your hands – it should become a circle around 15 cm in diameter and at least 7 cm thick at the widest point, curved on both sides.

3. Lean the ice against something sturdy (two logs, rocks or knives) and move your tinder (see page 121) or char cloth (page 124) into the burn point. This means you don't need to worry about holding the ice still. When your tinder catches after glowing and smoking, feed the tiny fire with birch bark and thin sticks before moving on to bigger fuel.

Swedish fire log

This is a one-log campfire that burns slowly and works well on the snow as the fire is off the ground. The flat top is ideal for a pot or pan and you can cook straight away without waiting for embers. And this fire looks really cool, too!

1. Saw a dry log to the desired size and use an axe to split it into halves, then quarters.

2. Stand the four pieces upright and fit them back together before fixing with wire or string toward the base and stuffing with tinder. Then, simply ignite! Your fire log should require no further intervention and will burn down slowly and steadily.

Vertical grill

As well as resting a pot or pan on the flat top of the fire log, I've also had fun grilling meat and vegetables vertically on skewers in the centre of the fire log after the four quarters have burnt down to embers. Try it!

45

Make a qulliq

A qulliq, or kudlik, is a traditional oil lamp as used by the Inuit. These lamps would have been made by burning seal and whale blubber or reindeer fat with a long wick of dried moss, cotton grass or cottony willow seed heads; useful combustibles in an environment without wood for fuel.

The qulliq would provide light and heat, even allowing clothes to be dried at the end of a long journey. They were burned in crescent-shaped carved stone bowls, later replaced by metal versions. These would traditionally be tended by the Inuit women; a symbol of their strength, tender love and care.

For an improvised qulliq, replace the animal fats with oil and use a fat wick of thick cotton.

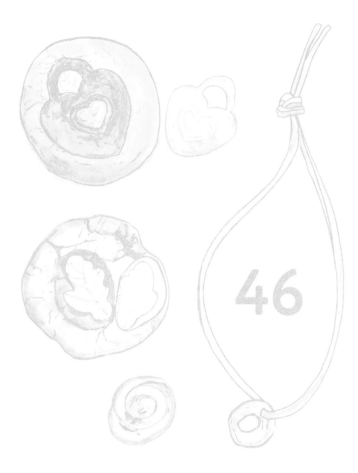

Pewter casting

Pewter is a tin alloy with a low melting point, which means that we can melt it over a fire. This opens up endless possibilities for casting and creating lasting treasures. This is a lovely activity to do all year round, but there is something about the combination of fire, metal and snow that adds an extra magic to this.

Equipment

- **Clay** Gather your own from the ground (with the landowner's permission) or buy natural clay

- **Pewter** Buy in shot or bar form or, even cheaper, buy off-cuts and scrap

- **Crucible** These can be shop-bought or you can create your own pouring lip on a metal ladle by striking one or both sides with a screwdriver and mallet

1. Create your mould or cast by pressing natural objects into the clay or by scraping and shaping the clay with sticks or modelling tools.

2. Heat the pewter in the crucible over the fire, before pouring it carefully into the cast. Wait until fully cooled before removing.

Other ideas

- Put a stick into the clay mould, which you can then remove from the final piece to create a hole for hanging or threading on a necklace.

- Drop hot pewter into a bucket of cold water for a satisfying hiss and a free-form shape. Could you interpret each person's shape and use it to 'tell their future'?

- Create a cast by carving into a cuttlefish bone with a nail or tent peg to give the final piece a slight yellowish tinge and a textured finish.

Fire stones

The simple combination of fire and stones can be an incredibly useful and potentially lifesaving tool out there in the extreme cold. Use non-sedimentary rocks gathered from high, dry places, as rocks with water inside could split or explode if heated. If you're not sure if your rocks are safe to heat, keep your distance from the fire as they warm up.

Cooking

Fry directly on a flat-topped hot stone, or boil water by dropping very hot smaller stones directly into your water.

Central heating

Put warm (not too hot!) fist-sized rocks in your sleeping bag at bedtime or put larger hot stones on a bed of branches or needles in your snow cave (see page 108). You will need multiple stones for bigger shelters and could even have a team at work on this, rotating roles (heating stones, swapping them in and out of the shelter and, of course, resting).

Injury and first aid

If one of your group is very cold and hypothermic, they can rub hot stones on their body – hands, armpits, neck and feet – to raise the body temperature. Warm rocks can also soothe strains, sprains and soreness; simply hold the stone against the affected area.

Ice fishing

Heat up a suitable rock and drop it onto the ice on a frozen lake. Watch it melt the ice, working its way down to create a perfect hole for ice fishing!

Evergreen tea

Rich in vitamin C, anti-inflammatories, antioxidants and herbal antivirals, this evergreen tea is nutritious and delicious! Species such as white pine, balsam, fir and spruce will work well. For details on identifying coniferous trees, turn to page 80.

Safety tip: Research the trees in your area, making sure you avoid toxic trees that look similar, such as yew. Avoid trees by the roadside and in highly polluted areas. There are also studies that suggest that pregnant women should avoid drinking pine tea.

1. Pick fresh green needles from the tip of the branch.

2. Wash and roughly chop the needles, then bash them lightly with the back of a knife to release the oils.

3. Steep the needles in boiled water off the fire for 10 minutes, then strain into cups. Sweeten with honey to taste.

Variation: Adding apple and cinnamon can be delicious too. How about experimenting with other fruits and spices?

49

Mexican hot chocolate

Bring some Mexican spice to a snowy scene with this delicious
hot chocolate. Made on the fire or hob, this one is sure to warm
the hearts and bellies of your people.

Serves 4
200 g dark Mexican chocolate
1 litre milk of your choice
1 cinnamon stick
1 chilli, halved and deseeded
pinch of salt
sugar or other sweetener, to taste

1. Place the chocolate, milk and cinnamon stick in a pan over medium heat and bring the mixture to a rolling boil. Once the chocolate has melted, remove the pan from the heat.

2. If you have a *molinillo* (traditional Mexican wooden whisk), roll the handle between your palms to get the chocolate nice and frothy.

3. For a spicy kick, add the chilli along with a pinch of salt and sweetener to taste.

Variation: On those cold nights out in the snow, how about adding a cheeky shot of mezcal, brandy or orange liqueur for a winter warmer?

50

Tell a fireside story

The snow is falling, you've got a roaring blaze going (see page 122) and now it's time to invite the voices of our ancestors into the fire circle …

As children we spent much of our time immersed in the world of myth: listening, creating, playing and telling. Humans are storytelling animals – it's how we make sense of the world. This is an ancient and primal act; our oral history and much of our shared culture was forged around the fire over thousands of years of gathering, sharing and listening. Really listening. We can still play our part in this, communing with those that came before us, as well as creating our own new stories.

The words on these pages were crafted alongside trusted and respected storytelling elder Jan Blake. These are not instructions, more a few humble invitations for those inspired to move more deeply into story-making and -telling. It goes without saying that you can also tell a story in any way that feels natural to you!

Find the story

- Sit with the elders of your community, seeking out the stories of your family and ancestors. Listen to and learn from tellers in all aspects of life – from an aunty round the dinner table to comedians, activists, indigenous leaders and professional storytellers. Explore books, new and old, as sacred sources of story.

  - Look at the stories of your culture; you will find treasures there. If you tell the story of another culture, do so with respect and gratitude.

  - Find a story that speaks to you. If it lights you up, it is more likely to light up your listeners! Or will you tell your own story?

    - Explore different versions of the same story and let the essential parts emerge – these will stay with you as you tell and retell it, finding your own place within the story.

Get cosy

- Create a fireside atmosphere that embodies the Danish concept of *hygge*: a cosy, warm environment of comfort and wellbeing. How about a cup of evergreen tea (see page 136) or Mexican hot chocolate (see page 138) and some treats to pass around?

  - Shelter your fire circle from the wind, dress up warm, gather plenty of blankets and skins (responsibly sourced or fake of course!) and lots of firewood.

Tell the story

- Don't learn it word for word, instead read or listen to it, then retell it. Let the story be told through you.

- Understand the 'characters' as real people, like family members. Avoid caricature and practise empathising with them, even the 'bad' ones – wonder where they are coming from and look for their humanity.

- Explore the story, and as you tell it over time, don't be afraid to go off on tangents that emerge in the telling.

- Take your time. Play with pacing and volume. Embrace the silent moments. These are golden.

- Let your life experiences, pain and wisdom infuse the story – be yourself, with all your nerves, wounds and power.

Finally, let the 'medicine' of the story work its magic on the listeners. Each of us will respond in different ways, resonating with parts of the story that hold personal meaning – it's not our job to tell others what that message is.

Most important of all, just enjoy telling the story and have fun.

Sources

Axe: fiskars.co.uk, gerbergear.com, gransforsbruk.com

Cooking and eating: kupilka.fi/en, msrgear.com, selfrelianceoutfitters.com

Cordage: paracord.eu, ropesandtwines.com

Ferro rod: lightmyfire.com, thewoodlandschoolltd.biz

Knife: benandloisorford.com, morakniv.se/en

Magnifying glasses: microscope.com

***Molinillo*:** mexgrocer.co.uk

Pewter: Buy new from trusted sellers or find people selling lead-free pewter scrap and off-cuts on second-hand sites. Make your own crucible (see page 133) or buy from monumenthandtools.co.uk / muddyfaces.co.uk

Saw: agawagear.com, silkysaws.com

Snow probe: mammut.com, msrgear.com

Snow shovel: fiskars.co.uk, msrgear.com

Tarpaulins: ddhammocks.com

Acknowledgements

The deepest of gratitude to the people who contributed directly and indirectly to this book, including Štěpán Přindiš, George Mayfield, Woody, James Brunt, Janine Gerhardt, Jan Blake, Andreas Kornevall, Johnny Juhl, Steve Le Say, Ready, Jon Cree, Marnie Rose and The Garden Classroom, Millie Darling, Mo Maynard, Adam Njenga, Abi Hopper, Emma Reveley, Susie Laker and the 2022 Nightingale Y6 tribe, Iain Isaacson, Olly Otley, Maria Sprostranova, Patrick Harrison, the Skrein and Smee clans, Sylvie, Pintxo, my Barcelona and Vinaixa family, Rhiannon and Kinship in Nature, Marina Robb and Circle of Life Rediscovery, Ellen Vellacott and the Super Roots crew.

Huge thanks and respect to Maria, whose illustrations lift the book far beyond my words, and Clare Double, Krissy Mallett and the team at Pavilion and HarperCollins for steering the project from inception to publication.

I honour our ancestors and indigenous people across the globe, whose knowledge and wisdom contributed to every page in this book.

Gratitude also to winter itself; its harsh and beguiling energy, its calm and quiet.

Finally, my eternal gratitude and love to Roxanne, Santi and Zia – my favourite adventurers.